YOUR KNOWLEDGE HAS VALUE

- We will publish your bachelor's and
  master's thesis, essays and papers

- Your own eBook and book -
  sold worldwide in all relevant shops

- Earn money with each sale

Upload your text at www.GRIN.com
and publish for free

Bibliographic information published by the German National Library:

The German National Library lists this publication in the National Bibliography; detailed bibliographic data are available on the Internet at http://dnb.dnb.de .

Imprint:

Copyright © 2016 GRIN Verlag
Print and binding: Books on Demand GmbH, Norderstedt Germany
ISBN: 9783668719170

This book at GRIN:

https://www.grin.com/document/427428

Rodney Mulelu, Marie Matee

Challenges Regarding Antiretroviral Treatment Programme Implementation in South Africa

GRIN Verlag

CHALLENGES REGARDING ANTIRETROVIRAL TREATMENT PROGRAMME IMPLEMENTATION IN SOUTH AFRICA

Author: Mr. Rodney Azwinndini Mulelu, Co-Author: Ms. Marie Matee

UNIVERSITY OF SOUTH AFRICA (UNISA), PRETORIA,
SOUTH AFRICA

Abstract: The researchers investigated the challenges regarding Antiretroviral Treatment (ART) programme implementation in South Africa.A qualitative method was used. The paper focused on the challenges that the South African Government is experiencing in implementing ART programme. Despite all the prevention strategies and plans that the National Department of Health of South Africa has, the HIV and AIDS is still one of greatest burden the country has ever experienced. In this paper, it is further indicated that we are all affected and it is important to work together towards UNAIDS Zero targets strategy for 2030, ZERO new HIV infections, ZERO discrimination and ZERO Aids related deaths. The research findings revealed two themes: social support and economic themes of the study. Factors reported influencing optimum adherence were also indicated and are captured in the discussion report of this paper.

Key word: Social support, Economic Support, Unemployment, Adherence, antiretroviral treatment, HIV and AIDS

1. INTRODUCTION

This paper investigates the challenges regarding Antiretroviral Treatment (ART) programme in South Africa. This section presents a background to the study undertaken. It provides information on the focus area of the study and it outlines the research problem and the ARV roll-out programme initiated by the South African National Department of Health. In addition, it also outlines the aims and objectives of the study and the research questions to be answered by the findings.

2. BACKGROUND TO THE PAPER

The South African government recognizes that the country has a very high burden of infection, with an estimated 6.4 million people living with HIV by the end of 2012[1]. In addition, South Africa's landscape of the national HIV epidemic has improved dramatically since 2001. South Africa has reduced new HIV infections by 41% between 2001 and 2011 [2]. However, as the South African government enters the final years of achieving the Millennium Development Goals in 2015, much remains to be done to reach the targets. A number of government interventions and strategies have been put in place over the years to reduce the HIV infection rates in the country. The programmes such as the HIV Counseling and Testing (HCT) awareness campaign established in 2010 [3], the increased call for condom use and distribution, HIV and sex education in all sectors of the society, and the medical male circumcision programmes were all aimed at reducing HIV infection.

The launch of the national HIV counseling and testing (HCT) campaign in April 2010 resulted in a remarkable increase in the number of people accessing testing [1]. It is further indicated by the HSRC (2012) that between 2008 and 2012, annual HIV testing increased from an estimated 19.9% to 37.5% among men, and from 28.7% to 52.6% among women.South Africa has the largest antiretroviral treatment (ART) roll-out programme in the world, achieving a 75% increase in HIV treatment services between 2009 and 2011 [4]. By October 2012, over two million people were receiving ART, surpassing the country's universal access target of 80% in accordance with the 2010 World Health Organization (WHO) treatment guidelines and offering treatment to people with a CD4 count under 350[5].

According to the National Department of Health (NDOH), the government is committed to ensuring universal access to antiretroviral therapy in order to improve the quality of lives and the country outcomes [6]. In addition, the national ART roll out programme for South Africa requires human resources, financial planning and monitoring and evaluation systems, as well as institutional capacities [7]. For ART to be able to attack and suppress viruses, patients must adhere to treatment for life. Not following instructions to take medication, collecting medication and not taking it and missing out on doses will lead to drug resistance and to the medication not to work [8]. Keeping patients on an ART programme needs commitment from all stakeholders, patients, family, programme managers, wellness clinic staff and political leaders. Furthermore, optimal adherence to ART has the potential for turning an HIV infection from an acute illness to a manageable chronic disease such as Diabetes and Hypertension.

The South African national ART programme was conceived to reduce and meet the high burden of infection that the country experiences [9]. South Africans hope for a good quality of life for people living with HIV and in need of ART when the Operational Plan for Comprehensive HIV and AIDS Care, Management and Treatment for South Africa was approved by the National Department of Health in 2003 [10]. The National Department of Health's antiretroviral treatment revised guidelines state that adults and adolescents are eligible for ART if they meet medical criteria and psychosocial considerations [11]. Furthermore, it means that the patients should meet these criteria to be able to start the treatment; the patient should be psychologically and medically ready to start ART, because ART is not an emergency. In addition the eligibility criteria for ART initiation have been revised to increase access to treatment, with effect from 1st January 2015. The medical criteria include children under the age of 5 years old, Adolescents and Adults with CD4 of < 500 cells/mm^3 or a diagnosis of WHO stage 3 and 4 diseases, irrespective of CD 4 count or WHO staging [11].

Furthermore, there is the psychosocial consideration that patients should have attended three or more scheduled visits, no active alcohol or substance abuse, disclosed of HIV status to at least one family member or joined support group and is able to visit the treatment centre on a regular basis [11].The strict adherence to antiretroviral therapy (ART) is the key to sustained HIV suppression, reduced risk of drug resistance, improved overall health, quality of life, and survival [12]. In addition, poor adherence to antiretroviral medication accelerates the development of drug resistant HIV, whereas without adequate adherence, antiretroviral agents are not maintained at sufficient concentrations to suppress HIV replication in infected cells and to lower the viral load. Furthermore, it is important to identify and address challenges to complete adherence when starting ART, because adherence is such an important part of the treatment [13]. It is clear that adherence is a cornerstone of a successful ART programme. It can promote retention of patients on ART for longer periods and the attainment of optimal health. Furthermore, The NDOH emphasizes the importance of adherence and suggests having a communication strategy that includes daily adherence reminders and re-adherence counseling at every clinic visit [14].

3. RESEARCH DESIGN

The research was conducted using the qualitative enquiry method to collect data. The goal of the paper was to evaluate and identify challenges of the ART implementation in South Africa. Given this goal, a qualitative method approach was called for. In this study, data were collected using interviews conducted with the people living with HIV who are on ART. Qualitative research is any data gathering technique that generates open-ended, narrative data [15].

3.1. Sampling strategy

Following the intensity sampling strategy as described [16], the researcher purposely recruited participants attending the wellness clinic in the facility. The participants for the face-to-face interviews were purposefully chosen as they present themselves at the wellness clinic. On a daily basis, clients present to a dietician consulting room for dietetic services. As they present to the Dietician (researcher), the researcher had an opportunity to request them to participate in the study. Those accepted to be part of the study were interviewed and five participants accepted to participate in the interviews.

4. DATA GENERATION STRATEGIES

Since a qualitative research design was used [17], the researcher followed one data generation strategy. It is discussed in greater detail below. The data collection is defined as the precise and systematic gathering of information relevant to the research purpose, objectives and questions [15]. The researcher was totally involved and able to interact with the participants.

5. DATA ANALYSIS

The interviews data, which was collected by means of tape recorder, was transcribed. The theoretical framework used in the research guided the conceptualization of the categories and themes in the data analysis. In this study, data analysis will start with listening to the tape recordings numerous times. The tape recorded interviews will be transcribed and translated to English. Similar patterns were extracted from the interview transcripts. The data were coded and analyzed manually. Themes were identified. In *thematic analysis*, the researcher is predominantly interested in the emergence of themes from the collected data.

6. DISCUSSION OF THE PAPER

6.1. HIV AND AIDS IN SOUTH AFRICA

The NDOH indicates that public sector antiretroviral provision has had a slow start in South Africa due to political denials despite a raging epidemic and a World Aids Conference that shed a light on the disparities of therapy access globally [9]. In addition, the South Africa has the largest number of people living with HIV (PLHIV), estimated at 6.4 million, and by far the largest number of people on antiretroviral therapy (ART) in the world: almost 2.5 million have started ART in South Africa with a total population of 54 million people [18]. Besides the logistical challenges that a programme of this scale presents, the expanding number of people requiring treatment poses significant resource challenges [4].

Following a troubled period, in which the South African government failed to come up with an appropriate treatment plan and which was accompanied by wide controversy and active campaigning within civil society, the government announced a gradual ART roll-out in 2003 [10]. South Africa's decision to start offering ART at facilities and committing itself to fighting the disease has brought some relief after its long-standing indecisiveness on this issue. In addition, there are between 1.6 and 2.0 million additional PLHIV eligible for ART initiation at CD4 T-cell counts of < 500 based on the 2015 WHO Antiretroviral Treatment Guidelines [19, 11].

In addition, it is indicates that between 300,000 and 500,000 people have become infected annually over the past decade, while the capacity of the National Department of Health system to start people on ART has expanded by only 20%[19]. It is difficult to overestimate the suffering that HIV has caused in South Africa. Not only the person living with HIV in South Africa - or in any other country - are affected, but also their families, friends and the wider community. However, after years of controversial AIDS treatment history and the refusal of the South African government to provide ART to people living with HIV in South Africa, the South African government finally announced its approval to start offering ART in 2004[2].

The treatment programmes have access to medication choices and monitoring that is in line with international guidelines, with the most recent improvements including fixed dosed combination (FDC) ART, including treatment for pregnant and lactating women and access to salvage regimens as reflected in the government HIV treatment guidelines [20]. According to HSRC, it is estimated that 12.2% of the population (6.4 million persons) were HIV positive, which is 1.2 million more PLHIV than in 2008 (10.6% or 5.2 million) [1].

6.2. HIV AND AIDS IN THE LIMPOPO PROVINCE

The National Antenatal Sentinel HIV and Syphilis Prevalence Survey (2012) shows that the Limpopo Province is seven-highest in terms of HIV prevalence in the country, after six other provinces such as KwaZulu-Natal, Mpumalanga, Free state, North West, Eastern Cape and Gauteng[20].

The report further indicates that Limpopo is amongst the few provinces that have shown an increase in the overall prevalence rate, that is, from 8.8% in 2010 to 9.2% in 2012. The Limpopo Province has five District Municipalities, the Capricorn, Vhembe, Mopani, Sekhukhune and Waterberg Districts. It borders on Gauteng, the North West, Mpumalanga Provinces, the Republic of Mozambique, Zimbabwe and Botswana. There is a lot of migration to and from these neighboring countries and provinces. Dr. Mabasa, the MEC for health in Limpopo province indicated that the Waterberg District (under which the chosen research site falls) continues to be the highest in new HIV infections in the Limpopo Province estimated at 30.3% at the end of 2012[21] This high rate of infections in the Waterberg District is attributed to high numbers of migration of contract workers in the Lephalale Local Municipality [21]. The ART programme implementation in the province was a call from National Department of Health for all hospitals in the country to be accredited to offer ART services [22].

6.3. SOUTH AFRICAN ANTIRETROVIRAL TREATMENT(ART) PROGRAMME

South Africa has the largest antiretroviral treatment (ART) programme in the world and has made significant strides in improving ART coverage [4]). The South African government, with the assistance of international funders and non-governmental organizations (NGOs), has managed to roll out a very effective ART programme since 2004 free of charge to the people living with HIV (PLHIV). At the end of 2009, an estimated 37% of infected people were receiving treatment for HIV [5]. In mid-2011, following the launch of the HIV Counseling and Testing (HCT) campaign by the SA National Department of Health in early 2010, it was announced that the number of people on ART had increased significantly from 923,000 in February 2010 to 1.4 million in May 2011[23]. However, HRSC (2012) reported that South Africa has reached a target of universal access to treatment with 2 million people initiated on ART by October 2012.

The introduction of the revised treatment guidelines which prescribed that HIV- infected individuals should be initiated on ART at a CD4 count of 500 and less has an impact on the attainment of this [11]. This directly promotes the strategy of using 'treatment as prevention' which espouses the notion that putting people on ART earlier will positively impact on the lowering of the overall number of new HIV infections and therefore result in fewer people needing treatment [20]. With the inability of various countries to control the HIV epidemic and the increasing number of new HIV infections, the treatment as a preventative measure for HIV infections was introduced. ART has been shown to be effective in preventing the acquisition of HIV infection when given as a prophylaxis before or after exposure to HIV to prevent transmission from HIV infected persons [11].

A recent shift proposed in the approach to curbing the scourge of HIV is the 'test and treat' strategy. This is the model of universal voluntary HIV testing with immediate commencement of ART for those diagnosed HIV positive [11]. Furthermore, it is proposed as another form of using treatment as a preventative measure for HIV infection. The model is likely; however, to have its own challenges relating to adherence and needs to be explored further as a prophylactic model for HIV. The provision of ART to persons living with HIV and requiring treatment in South Africa has brought hope to people who otherwise would not have survived. In addition, it has led to improved quality of life, reduced morbidity, mortality and the number of deaths resulting from AIDS-related illnesses and has generally prolonged the lives of many people living with the virus [24]. In addition, The UNAIDS reports that there is a noted decline in the number of AIDS-related deaths by at least 24% since 2005 and in 2011 there were 1.7 million deaths reported [4]. Despite these advantages, though, the ARV roll-out programme has had its fair share of challenges such as non-adherence or poor adherence to ART and the resultant high incidence of patients being lost to follow up or dropping out of the programme. Although public sector programmes for providing ART in the Sub-Saharan region have matured tremendously, a high rate of patient attrition is still evident. They further note that while much of the attrition is due to mortality, a loss to follow up or disappearance of patients from treatment, with no apparent reason, is very common[25].

Approximately 59% had disappeared from the system and could not be accounted for in the first year of being initiated on treatment [26]. The international studies such as the one conducted by Atav, Senir and Darling reported that the knowledge of HIV and AIDS patients and the public in general regarding HIV and AIDS policies, such as the ART policy of South Africa, is still not sufficient to form a basis for change in behaviour among the at-risk groups and people living with HIV [27]. According to NDOH, the ART policy of South Africa does not address the issue of poverty among people living with HIV until their CD4 count is below 500, which is when the patient qualifies for a disability grant provided by the Social Department of South Africa. Furthermore, the policy indicates that people who are HIV positive and have a CD4 count of less than 500 or have reached stage 4 of the WHO clinical staging, qualify for a disability grant and should be started on ARV immediately. Obviously the protocol of initiation of ARV has to be followed [11]. This leads to people intentionally having sex with someone who is HIV positive, or they stop taking the ARV drugs so that they can qualify, or so that the disability grant should not be stopped. For example, high-risk groups such as commercial sex workers, due to poverty and the need to survive will engage in unprotected sex in order to earn money [28]. High-risk sexual engagement such as sex without using a condom yields more customers and a higher income. Change in behaviour is dependent on what the individual perceives as benefits as opposed to what the individual perceives as risks or threats [28].

Patients are more inclined to change their sexual behaviour after the initiation of ART and prevention counseling [29]. Sexual desire changes over time, with many in their research reporting diminished desire at 3 to 6 months of taking ART as compared to 18 to 24 months of use. Some patients reported that they feared re-infection or infecting others, or that engaging in sex would awaken the virus and weaken them, and that they might die [30]. Hoang, Ding and Groce researches on female African-Americans indicate that most respondents believe that they will pray to God, or according to their culture, they believe that their husbands or boyfriends are being faithful [31]. It is evident in most of these studies that patients living with HIV and AIDS lack knowledge about antiretroviral treatment [32].

7. LEGAL FRAMEWORK INFORMING SOUTH AFRICAN ART PROGRAMME

According to Sithole (2013), there are two important strategic documents informing the NDOH ART programme in South Africa: namely, the National Strategic Plan (NSP) and the National Antiretroviral Treatment Guidelines to assist with the implementation of the programme. These two documents are discussed briefly below.

7.1. National Strategic Plan (NSP) of 2012-2016

The NSP reflects the progress made in achieving a clearer understanding of the challenges posed by these epidemics and the increasing unity of purpose among all the stakeholders, who are driven by a shared vision to attain the highest impact of policies towards long-term vision of zero new HIV and TB infections [15]. NDOH indicates that ART expansion programme has resulted in an increase in ART facilities countrywide to about 2 552 currently and more people accessing treatment. The NSP focuses specifically on expanding the quality and reach of health and wellness and is geared to addressing the gaps identified in the previous NSP of 2007-2011[15]. The gaps identified include inadequate co-ordination of the public sector, private sector and non-government sector responses, the weak governance and co-ordination structures of SANAC (from ward to national level), the lack of robust monitoring and evaluation of the NSP, the failure to ensure a truly multi-sectoral and integrated response, weak focus on human rights and justice, and the lack of a comprehensive and integrated approach to HIV and TB prevention. The NSP suggests that these gaps be addressed through annual HIV testing for everyone in South Africa, initiation of every HIV-infected individual on ART when their CD4 count is 500 and less and the strengthening of adherence counseling programmes to ensure retention in care [11].

7.2. National Antiretroviral Treatment Guidelines 2004 and 2015

The main purpose of these guidelines is to improve the clinical outcomes of people living with HIV, to reduce morbidity due to TB/HIV co-infection, to reduce HIV incidence and to avert AIDS-related deaths in the most cost-efficient manner by ensuring that people living with HIV start with the right therapy at the right time. The National Antiretroviral Guidelines also seek to ensure timely HIV diagnosis, management, treatment and initiation of ARVs for treatment for all eligible populations to achieve best health outcomes in the most cost-efficient manner [11]. The NDOH focused on defining adherence and formulating strategies to support and improve adherence to ART, clearly defining roles and responsibilities of the Health Care Team [24].The NDOH indicated that the Honourable President Jacob Zuma announced new key interventions to improve antiretroviral treatment (ART) access to special groups such as HIV positive infants, pregnant women and TB and HIV co-infection [33]. This announcement resulted in more than 2.6 million people being initiated on ART by mid-2014 [11]. Furthermore, in 2013, the fixed-dose combination pill (FDC) was introduced, made up of the regular three drugs used in the first-line regimen to improve adherence and retention [20].

In addition, in 2014, the South African Minister of Health, Dr. Aaron Motsoaledi announced that the threshold for initiation of ART will rise to CD4 count ≤500 cells/µl and that the PMTCT programme will now adopt the B+ approach, which entitles every pregnant and breastfeeding woman to lifelong ART regardless of CD4 count or clinical stagin[9].In the new guidelines effected by NDOH, the approach focused on providing the patient with a comprehensive treatment plan and on-going monitoring to ensure adherence[11]. In January 2015, the treatment guidelines were reviewed and a new set was introduced which re-emphasized the importance of adherence and of socio– economic support so as to ensure positive treatment outcomes for patients and the identification of issues that impact on optimal adherence.

There was also a shift in the approach used as the NSP emphasizes negotiating the treatment plan and allowing the person to commit to the plan. They further identified main factors that have an impact on adherence and these are personal and environmental factors [11].

Furthermore, these successes are underpinned by a range of interrelated, evidence-guided strategic and operational plans, monitoring initiatives, and policies and guidelines, paving the way for South Africa to attain its objective of reducing mortality from HIV and TB. However, there is a need, to strengthen the focus on adolescents and prevent them from acquiring HIV in line with Department of Health 2020 strategy [34]. These new guidelines will assist in providing the necessary guidance towards improved management of HIV across different populations in South Africa

8. CHALLENGES REGARDING ANTIRETROVIRAL TREATMENT PROGRAMME IMPLEMENTATION IN SOUTH AFRICA

Despite the success of the programme, it is not free of challenges. Some of the identified challenges that may have an impact on adherence to ART in the long term are discussed below. The success of the ART programme is determined by early identification of infection in PLHIV, rapid and appropriate starting of ART, high levels of viral suppression as well as the management of toxicities [24]. In addition, the current status of the ART programme and its implementation in South Africa is encouraging but significant challenges remain and some are discussed in the following subsections.

8.1. Linking and retention of patients to care

Retention of patients to care is referred to as client's continued engagement in health services and captures the whole continuum of HIV care from enrolment in care to discharge or death of the client. In addition, retention in care to antiretroviral treatment (ART) is a critical element of HIV care and associated with optimal individual, public health outcomes and cost effectiveness [5].

The challenges facing the ART programme in South Africa is the ability to initiate and retain people living with HIV on their treatment regimens, according to Stricker, Fox and Gill [35]. Furthermore, there is strong evidence that many patients with chronic illnesses such as asthma, hypertension, diabetes and those living with HIV have difficulty adhering to their recommended regimens. The outcome of HCT is only successful if those who are HIV – negative are supported to reduce their risk of acquiring HIV and also those HIV positive are successfully linked to the continuum of care[11]. Despite the significance of retention in care, patients living with HIV are still finding it difficult to follow the recommended behaviour by the health care providers. In addition, inadequate retention leads to decreased health outcomes such as morbidity, mortality, and drug resistances, risk of transmission, increased costs and lower productivity [35].

8.2. High levels of loss to follow up (LTFU) to Antiretroviral treatment

Loss to follow up (LTFU) represents clients who disengaged from care at any stage of the continuum of care. Despite the efforts of the health providers, for various reasons, there is still a significant loss of patients from the time that people are diagnosed with HIV to the first assessment of ART eligibility [35]. According to NDOH the treatment cascade shows two main leakages: people lost between a positive HIV, CD4 test, and those lost between CD4 test and the return visit for CD4 test result [11]. Challenges facing people living with HIV include having to travel long distances to the clinic, long waiting times at the clinic, clinic staff shortages, inability to take time off work, lack of full understanding of the treatment plan and fear of stigma and discrimination.

However, if these patients are effectively linked to prevention, treatment and care services, HCT enables those being tested to make positive health –related decisions. Despite these gains, the challenge of late presenters at the facilities continues to occur throughout the world including South Africa, with the resultant ongoing burden of opportunistic infections including TB [19]. In addition, the pace of the implementation of the ART programme in the provinces is still impeded by the human resources crisis that is being experienced by the Department of Health. The lack of trained professionals, because of the difficulty of attracting, training and retaining of health care workers, remains a formidable challenge for the Department of Health. Without addressing the human resources issue, which includes poor working conditions, low salaries, unsatisfactory career pathing and lack of incentives, it will be difficult to increase the pace of ART implementation in the province [36].

8.3. Challenges regarding adhering to antiretroviral treatment

The biggest challenge facing the ART programme in Sub-Saharan Africa and South Africa is the ability to initiate and retain people living with HIV on the treatment regimens [14]. It was indicated that although public sector programmes in the Sub-Saharan region have matured, a high rate of patient attrition is still evident. They further indicate that while much of the attrition is due to mortality, loss to follow up or disappearance of patients from treatment with no apparent reason is very common[25].

ART programmes in Africa only managed to retain approximately 60% - 70% of patients on therapy three years after initiation on therapy [37]. Furthermore, the study regarding the barriers to adhering to ART found that among patients living with HIV, many reported that they forgot the sequence of the treatment. In addition, they could not take the medication in front of their family members or friends because they did not know about their HIV status. Furthermore, some reported that the medication was too complex and they also lacked knowledge about the treatment [38]. In addition, the similar study conducted in Botswana also reported barriers such as stigma, discrimination, migration and side effects of ART as key challenges to adhering to and complying with ART. In this study, the cost of ARV drugs was found to be a serious challenge [39]. Furthermore, the trust between health care providers and the patient are crucial for successful implementation of an ART programme. The study was conducted in the Connecticut Department of Corrections in the United States of America, and they looked into trust between the health care providers and prison inmates and the acceptance and adherence to antiretroviral medication. It is therefore essential that adherence should be improved by intervention designed to address the barriers. People living with HIV must be given accurate information for them to understand what it is that they as patients need to do [39].

Stigma; discrimination and the issue of privacy are some of the barriers to the success of ART implementation, especially regarding the problem of disclosure to family members and friends. HIV and AIDS patients need to disclose their status to one person, a member of the family or a friend, who will support them [40]. In addition, it is indicated that privacy is also considered another impediment to treatment and adherence, because in choosing to keep their positive status a secret, patients might not get treatment. It is also indicated that depression and psychiatric conditions such as depression and stress especially among adolescents, occur as a result of non-adherence to the treatment. It has been noted that adolescent patients taking ARV are unable to cope with depression or stress and this causes them to miss treatments [40]. Participants with partners, such as people in heterosexual relationships, were associated with good adherence, that is, taking their ARV prescription as prescribed, while support from friends and other family members was not as significant[41]. Their findings also indicate that partners who were both HIV positive and both taking ARV supported each other more in taking ARV than friends and other members of family or relatives. A study conducted by Berg indicates that there are main barriers to ART success, not limited to the following: forgetfulness, the social and physical environment, the complexity of the regimen, side effects of the medication, inadequate knowledge on the part of the patient about ART, such as its benefits and side effects [42]. However, Berg points out that these barriers may be overcome by providing patients with sufficient health care information in terms of the benefit of ART, comprehensive pre-treatment, post-treatment and on-going counseling at each visit to the facility[42,43].

9. FACTORS FOR POOR ADHERENCE

Factors that may have either positive or a negative impact on an individual adherence behavior may be divided into three categories, namely patient related, regimen related and disease related (as cited by Mathebula)[44,45]. The major impact on adherence lies with the individuals on therapy and their interactions with their families, communities and care givers. The more favorable these relationships, the more likely the individuals are to remain adherent over time. The value of an individual accepting the HIV status and being properly prepared for therapy should not be underestimated. Many people simply just forget to take the medication. The use of devices such as pill boxes, cell phones and diaries to remind them to take their medication can be useful in this case[24].

9.1. **Socio-economic factors**

Social and economic factors may combine to yield poor adherence outcomes in South Africa. Many people are living in poverty in South Africa and have limited or no financial resources to meet treatment requirements [46]. These circumstances may lead to poor adherence outcomes. On the other hand, however, it is stated that socio-economic factors are not consistently predictive of adherence [47]. They assert that in studies carried out in India and South Africa, no association was found between adherence and the patient's economic status. A number of investigators relate the costs of accessing treatment at the level of the individual and his or her family to poor adherence outcomes [47].

For example, with regards cost such as transport and waiting times as important barriers in addition to stigma, family pressures and religious beliefs[48]. In addition, in a review of studies conducted in Tanzania, Uganda and Botswana, conclude that although patients are highly motivated to adhere to their medication as prescribed, constraints such as transport costs, user fees, long waiting times, hunger, stigma and discrimination, lack of social support, side effects and lack of appropriate counseling all undermine such intentions [49]. It is also confirmed that costs such as payment for transportation to and from clinics serve as a deterrent to ART adherence [50]. They further state that the lack of adequate food has also been associated with poor adherence to ART. Socio-economic factors reported to have a significant impact on adherence are poor socio-economic status, poverty, illiteracy, low levels of education, unemployment, lack of effective social support networks, unstable living conditions, and long distance from treatment centers [51].

However, state that educational levels, literacy, income and housing status are not predictive of adherence [47]. In addition, it is reported that poverty can influence adherence because access to the financial means to travel to and from the ART clinic, to pay for child-minders during a parent's absence to access treatment and to attend to several, competing needs and responsibilities all come into play [52]. For the poor and the unemployed, lack of financial resources to pay for the kind of food stuffs required to be taken alongside medication may be of particular concern [53]. According to NDOH, poverty is one of the major contributors to poor health and treatment outcomes through food insecurity in as far as HIV and TB acquisition and treatment adherence are concerned. The SA government, in order to support adherence through its Social Development Department, has introduced a chronic disease grant assistance for those HIV infected who cannot support themselves and their families [20].

The qualification criterion for the grant is that an HIV-infected person must be too sick to work and when their health improves the grant is terminated [54]. From a study conducted in the Cape Peninsula, South Africa, with 29 patients, it was found that in order to keep their CD4 counts low, some manipulated the treatment so that their viral loads remained high[55]. Furthermore, the commonly cited reason for non-adherence is the fear of losing one's grant and that this results in some patients 'managing' their CD4 counts by skipping some doses. This ensures that the viral load is not completely suppressed and this enables them to continue receiving the grant as long as there is no improvement in the CD4 count. This can have devastating effects in terms of developing treatment resistance and treatment failure in the long run and the patient may end up with fewer treatment options (Sithole 2013). As far as social factors are concerned, the WHO suggests that positive attitudes of the community towards people living with HIV play a significant role in influencing adherence [56].

The WHO indicates that people living with HIV and on ART and receiving community-based support have been reported to be less likely to be lost to follow-up [56]. It is indicated that of the 19,668 patients studied in a cohort, only 6% of that receiving community-based adherence support was lost to follow-up compared with 9% of those not receiving any community-based adherence support [57]. The findings of a study conducted in five different sites in KwaZulu-Natal on the experience of patients on antiretroviral therapy suggest that challenges such as the availability of a treatment supporter, adequate household income and side-effects are all determinants of adherence, non-adherence and/or ultimate drop-out from treatment. Another important social factor, which is associated with adherence, is the support of significant others in a patient's life as they provide support and understanding for the patient [58]. In addition, it is confirmed that the presence of social support systems such as supportive family members, friends and treatment supporters has been consistently associated with good adherence to treatment [47].

9.2. General patient-related factors

Other factors identified as impacting on adherence are as follows:

- Treatment fatigue and pill burden. According to NDOH, having to take a number of pills as a result of co-infection with other illnesses such as TB and hypertension can result in fatigue and pill burden. This can result in patients adhering sub-optimally to their treatment. Some patients may also develop 'treatment or pill fatigue' after being on treatment for a long time and they may suddenly find it difficult to adhere to their recommended treatment regimens. Several research studies have reportedly shown adherence declining over time for people on long-term treatment and even with people who have been very successful in taking their medication [57].
- Lack of disclosure, fear of stigma and discrimination [46]. People living with HIV are often subjected to a great deal of stigma and discrimination which may result in their not being willing to disclose their HIV status for fear of being discriminated against.
- Cultural and religious beliefs [46] play a very significant role in shaping and determining beliefs and values which may directly influence attitudes to treatment and adherence.
- Forgetfulness and a lack of planning [3]. These factors may be manifest when the patient attends social functions on weekends or is away from home and no provision has been made in terms of having adequate treatment available.
- Lack of adequate financial resources and food insecurity [3] is cited as one of the major problems experienced in South Africa. This may result in the lack of money for transport to attend clinic or lack of money for food to be taken with the medication and this can affect adherence.

The individual's motivation, determination and an ability to manage his or her illness, available resources and support systems, understanding of possible consequences of non-adherence and expectation of positive treatment outcomes all interact to impact on adherence [24].

10. DISCUSSION OF THE FINDINGS

10.1. THEMES EMERGED FROM THE INTERVIEWS

10.1.1. Theme 1: Social support for PLHIV on Antiretroviral treatment

This theme describes the support structures available for PLHIV on ART and the importance of having these when coping with the demands of the ART programme. Participants mentioned a variety of support needs, including: emotional support, food, family support, and support from health professionals. In this study, it was found out that families were placed at the forefront of social, emotional, physical and economic support for ART and care. Relatives are often sources of information and encouragement on advising other members to go for HIV testing, including where to go and where to access support for HIV related care and ARVs.

"My mother and my children know about my status and support me a lot. My neighbours know as well. The support group was there, where even the social workers used to come and talk to us but now is no longer there." (Participant 1). "I think the home is an important support system, people who are close to you need to disclose to make it easy, even if you forget the pills someone at home will remind you by asking if you've taken your pills." (Participant 2) "I knew my status in 2005 and my health status was not good, my sister kept on saying I must take ARVs...I asked my sister to come and sleep with me...she kept on reminding me not to forget and I told her I'll never forget." (Participant 3) "I have no problems; my daughter supports me and my parents are a warm, my children make sure I take pills at the correct time, they ask me if I've taken pills, they even pack for me the ones I take when I'm at work... my daughter is with me through all hardships." (Participant 4).

Community health workers (CHWs) are significant people in providing support for PLHIV. Their roles include home visiting for further support regarding adherence to treatment (including treatment for opportunistic infections), assisting with the various difficulties of disclosure and making arrangements for food parcels if necessary[14]. *"There are CHWs who do home visits so she was referred to come to my house. I was able to visit the clinic again. She motivated me because I wanted to desert taking treatment but she encouraged me to continue with my treatment."* (Participant 1) Almost all participants highlighted the support they got from the nurses and doctors who were rendering treatment, care and support in the ART programme. *"Yes, sometimes I used to not sleep at night. I went to the clinic and talk to (the) nurses and they told me to continue taking the treatment. They are doing a wonderful job and I am happy with them."* (Participant 2) *"I do talk to them if I have a problem and I am happy with them."* (Participant 3). *"Yes, I do, and they are doing ok and I am happy with them. If you have a problem, the sisters have love, they don't talk bad about us, they are nice to us."* (Participant 4). *"Yes, I talk to them, they assist me a lot. I call them teachers."* (Participant 5)

This theme is an important factor in supporting adherence, and it was mentioned by most of the participants. However, ARV patients are usually shocked when they first realize they have contracted HIV, and often find it difficult to disclose to family members and community support groups and so do not receive the social support they deserve and it is shown by their responses below. *"I did not want people to see me taking treatment. I used to take my treatment in town because of that. I am free now because we are many here who are taking treatment."* *(Participant 2)*

In contrast, other participants indicated that they have a support group and used to attend on Thursdays and Fridays. *"There are a lot of support groups here. We attend Thursdays and Fridays. My family supports me a lot. I try to assist other people. Other people believe when you talk about yourself, your experiences, and what you have went through. I told my mother, I told my brother, they all support me ..."* (Participant 3)

People affected by and infected with HIV are often subjected to discrimination and rejection by their friends, family, sexual partners and work colleagues. Even though they may be motivated to adhere to their ART for personal reasons, lack of or poor support from their loved ones, or feelings of being discriminated against or stigmatised as a result of being HIV positive may negatively impact on adherence. In this study, the findings are that all five participants reported that they have disclosed to family members especially the sister and the children. This is what they had to say:

"Yes,I have disclosed to my daughter who was born in 1993. I also told my sister. She accepted and did not have a problem." *(Participant 2)"Yes, I did disclose to my sister, then my children. My sister did not have any problem and she said I must just take treatment."* (Participant 3). *"Yes, I spoke to my brother and my father. My father was not happy but he took me to town and we talk about it. After that he was ok with that."* (Participant 4). *"Yes, I told my mother and my sisters. They told me is part of life. I don't have to think otherwise and they told me you will live like others."* (Participant 5). It is evident from this study that almost all the participants interviewed, reported that they received some social support from families and some engaged in community groups activities. WHO suggests that positive attitudes of the community towards people living with HIV play a significant role in influencing adherence? Adults on ART and receiving community based support [56]. It was found out that of the 19,668 patients only 6% of those receiving community based adherence support were lost to follow up as compared to 9% of those not receiving community-based adherence supports [57]. In addition, the presence of social support systems such supportive family members, friends and treatment supporters have been consistently associated with good adherence to treatment. In this study, the findings correlate with the statements given and shown in the responses above of all the participants interviewed [47].

10.1.2. Theme 2: Unemployment and economic support of PLHIV on Antiretroviral treatment

This theme describes the unemployment and economic factors that support the participants. Questions such as whether they are earning an income and whether there are any existing economic services helpful for PLHIV were asked, to understand the participants economic support factors. Furthermore, unemployment plays an important role in whether patients are able to adhere to their treatment regimens or not. It relates to factors such as transport and food costs that patients incur when they attend clinic in order to take their medication well[3].

Two of the five participants interviewed were unemployed and the other three participants were employed temporary, working piece jobs as they call it. None of the women interviewed were in a relationship due to issues of lack of trust. Financial difficulties were reported almost by all participants. Their responses are indicated below: *"Many people are weak and don't have power to work, I think giving them a grant or food will be ok. Many people struggle because of not having food to eat."* (Participant 1). *"I think the government should give us a grant for the disease." (Participant 3). "About the grant, we must be assisted. It is difficult if the tablets are finished and you don't have money for transport to the hospital." (Participant 5)*

Many patients experienced financial problems which meant they encountered problems in obtaining appropriate food and paying for transportation to the nearest facility to collect their medication [58]. It is indicated that patients struggled with transport and user fees, long waiting times, lack of food, side effects from medication and stigma [49]. The findings in this study show that three (60%) were employed, even though in 'piece jobs' and two (40%) were unemployed and dependent on the social grant. All participants interviewed struggled financially. Furthermore, a study conducted also found that scarcity of resources was a constant problem for adherence and patients had to beg, borrow or otherwise struggle to fund for transport to obtain their monthly medication [60].

11. CONCLUSION

In this paper HIV and AIDS in South Africa, Limpopo province and the South African ART programme has been discussed and presented. The paper focused on the challenges of the ART programme implementation in South Africa. Factors of poor adherence as some of the challenges have been presented. The findings of this paper were grouped into two themes and were presented in detail in this paper. Despite all the prevention strategies and plans that the National Department of Health of South Africa has, the HIV and AIDS is still one of greatest burden the country has ever experienced. In this paper, it is further indicated that we are all affected and it is important to work together towards UNAIDS Zero targets strategy for 2030, ZERO new HIV infections, ZERO discrimination and ZERO Aids related deaths.

REFERENCES

1. HSRC. (2012). Plenary Session 3, 20 June 2013. HIV/AIDS in South Africa: At last the glass is half full. Available at: http://www.hsrc.ac.za/en/media-briefs/hiv-aids-stis-and-tb/plenary-session-3-20-june-2013-hiv-aids-in-south-africa-at-last-the-glass-is-half-full
2. UNAIDS. (2012). Report on Global AIDS Epidemic: Geneva. Available at: http://www.unaids.org/en/media/unaids/documents (Accessed 9 May 2014).
3. NDOH. South African National Department of Health. (2010). *National Antenatal Sentinel HIV and Syphilis Prevalence Survey in South Africa*. Pretoria: NDOH.
4. UNAIDS. (2012). *The global AIDS epidemic – key facts*. Available at http://www.unaids.org. (Accessed 30 May 2014).
5. WHO. World Health Organization. (2010). *Guidelines on antiretroviral treatment. 2010*. Available at http://www.who.int/. (Accessed on 15 August 2014).
6. Jones, P.S. (2009). *AIDS treatment and human rights in context*. New York: Palgrave Macmillan.
7. Saleh, A.M. (2012). Exploring factors influencing adherence to antiretroviral therapy among adult people living with HIV on antiretroviral therapy in Zanzibar (Tanzania). A Case of Mnazi Mmoja Hospital Care and treatment Centre. Kit Royal Tropical Institute. Tanzania.
8. Abdool Karim, S.S. and Abdool Karim, Q. (2010). HIV/AIDS in South Africa. Second edition. Cambridge.
9. NDOH, South African National Department of Health. (2014). *Clinical guidelines for the management of HIV & AIDS in adolescents and Adults*. Pretoria: NDOH
10. NDOH. South African National Department of Health. (2003). *Operational plan for comprehensive HIV and AIDS care, management and treatment for South Africa*. Pretoria: NDOH.
11. NDOH. South African National Department of Health. (2015). *Clinical guidelines for the management of HIV & Aids in adolescents and adults*. Pretoria: NDOH.
12. Ross, E.L., Weinstein, M.C. and Schackman, B.R. (2015). Clinical role and cost effectiveness of long acting antiretroviral treatment. *Clinical Infectious Diseases, 57*(5). 1159.
13. Lyimo, R.A., De Bruin, M. and Van der Boogaard, J. (2012). Determinants of antiretroviral therapy adherence in northern Tanzania: a comprehensive picture from the patient perspective. *BMC Public Health*, 12(1). Biomed Central Ltd. 716.
14. NDOH. South African National Department of Health. (2011). *National strategic plan for HIV and AIDS, STIs and TB (2012-2016)*. Pretoria: NDOH.
15. Creswell, J.W. (2013). Research design: Qualitative, quantitative, and mixed methods approaches. Thousand Oaks. CA:Sage publications
16. Collins, K.M.T., Onwuegbuzie, A.J. and Jiao, Q.G. (2006). Prevalence of mixed methods sampling designs in social science research. Evaluation and Research in Education 19(2): 83-101.
17. Corbin, J. and Strauss A. (2014). Basics of qualitative research: Techniques and procedures for developing grounded theory. Thousand Oaks. CA: Sage publications.
18. Lalloo, U.G. and Pillay S. (2008). Managing TB and HIV in Sub-Saharan Africa. Current HIV/AIDS Reports, 5(3). Springer Publishers. 132-139.
19. WHO. World Health Organisation. (2013). The World Health Report 2013: Primary health care (now more than ever).
20. NDOH, South African National Department of Health. (2013). *Clinical guidelines for the management of HIV & AIDS in adolescents and Adults*. Pretoria: NDOH

21. Mabasa, N. (2013). Available at: <http://www.polity.org.za/article/sa-norman-mabasa-address-by-the-limpopo-health-mec-at-the-provincial-budget-vote-for-the-department-of-health-limpopo-legislative-chamber-lebowakgomo-23042013-2013-04-23>
22. NDOH. South African National Department of Health. (2004). *National Antiretroviral Treatment Guidelines*. Pretoria: NDOH.
23. AVERT. (2011). *HIV and AIDS treatment*. Available at http://www.avert.org/aidssouthafrica.html (Accessed on 20 May 2014).
24. Sithole, B.M. (2013). Factors that influence treatment adherence for people living with HIV and accessing antiretroviral therapy in rural communities in Mpumalanga. University of South Africa. Pretoria.

25. Rosen, S., and Ketlhapile, M. (2010). Cost of using a patient tracer to reduce loss to follow-up and ascertain patient status in a large antiretroviral therapy program in Johannesburg, South Africa. *Tropical Medicine and International Health* 15(1): 98-104.

26. Fox, M.P. and Rosen, S. (2015). Adult Patients on Antiretroviral Therapy in Low-and Middle-Income Countries: Systematic Review and Meta-analysis 2008-2013, *JAIDS,* 98-108.

27. Atav, S., Senir, D. and Darling, R. (2015). Turkish and American Undergraduate Students' Attitudes Toward HIV/AIDS Patients: A Comparative Study,. *Nursing Forum*

28. Sileo, K., Kintu, M. and Chanes-Mora, P. (2015). Such Behaviors Are Not in My Home Village, I Got Them Here": A Qualitative Study of the Influence of Contextual Factors on Alcohol and HIV Risk Behaviors in a Fishing Community on Lake Victoria, Uganda, *AIDS and Behavior*, Springer Publishers. New York.

29. Jean, K., Moh, R. and Danel, D. (2014). Decrease in sexual risk behaviours after early initiation of antiretroviral therapy: a 24-month prospective study in Cote d'Ivoire, *Journal of the International AIDS Society*. 17(1). 20-35.

30. Wamoyi, J., Mbonye, M., Seeley, J. and Bimungi, J. (2011). Changes in sexual desires and behaviours of people living with HIV after initiation of antiretroviral treatment: Implications for HIV prevention and health promotion. BMC public health Vol11. Available at <http://bmcpublichealth.biomedcentral.com/articles/10.1186/1471-2458-11-633> (Accessed 30 June 2013).

31. Hoang, D., Ding, A.T. and Groce, N. (2015). Knowledge and Perceptions of HIV-Infected Patients Regarding HIV Transmission and Treatment in Ho-Chi-Minh City, Vietnam. *Asia-Pacific Journal of Public Health*, 27(2). Sage Publications. 746-757.

32. Medley, A., Bachanas, P. and Grillo, M. (2015). Integrating prevention intervention for people living with HIV into care and treatment programs. A Systematic Review of Evidence, JAIDS.

33. NDOH. South African National Department of Health. (2009). *National Antenatal Sentinel HIV and Syphilis Prevalence Survey in South Africa*. Pretoria: NDOH.

34. Vun, M.C. (2014). Achieving universal access and moving towards elimination of new HIV infections in Cambodia. *Journal of the International AIDS Society*.17(1). 18905.

35. Stricker, M., Fox, S. and Baggaley, R. (2014). Retention in care and adherence to ART are critical elements of HIV care interventions. *AIDS and Behaviour*, 18(5). 465-475.

36. Ford, P. (2014). Nurse initiated and managed anti-retroviral treatment: An ethical and legal analysis in South Africa. Wits University, Johannesburg. p24. Available at http://hdl.handle.net/10539/14413. (Accessed 4 November 2015).

37. Bucciardini, R., Fragola, V. and Abegaz, T. (2015). in: Care of Adult HIV Patients Initiating Antiretroviral Therapy in Tigray, Ethiopia: A Prospective Observational Cohort Study,. *PLoS one*, 136-146.

38. Gengerg, B.L., Lee, Y. and Rodgers, W. (2014). Four types of barriers of adherence to antiretroviral therapy are associated with decreased adherence over time. *AIDS and Behaviour*, 85-92.

39. Van Boekel, L.C. (2013). Stigma among health professionals towards patients with substance use disorders and its consequences for healthcare delivery: systematic review. *Drug and Alcohol Dependence*, 131(1). 23-35.

40. Wasti, S.P., Simkhada, P., Randall J., Freeman J.V. and van Teijlingen E. (2012). Factors influencing adherence to antiretroviral treatment in Nepal: A mixed method study. PLoS one. 5(7) 35547

41. Warner, L.M.R. (2013). Interactive effects of social support and social conflict on medication adherence in multimorbid older adults. *Social Science & Medicine*, 87(1). 23-30.

42. Berg, H.E. (2015). The Effect of Familial Support, Socioeconomic Status and Stigma on Adherence to Antiretroviral Therapy Among Women in Hartford CT: A Qualitative Approach.

43. Nchendia, A.I. (2012). *Barriers to adherence to antiretroviral treatment in a regional hospital in Vredenburg, Western Cape*. Cape Town: University of the Western Cape.

44. Maartens G., Cotton M., Wilson D., Venter F., Meyers T. and Bekker L. (2012): The handbook of HIV medicine, 3rd Edition. Oxford University Press, Southern Africa.

45. Mathebula, T.J. (2014). *Reasons for Default Follow-Up of Antiretroviral Treatment at Thekganang ARV Clinic*. Pretoria: University of Pretoria.

46. Kagee, A. (2006). Adherence to antiretroviral treatment in the context of national rollout in South Africa: Defining a research agenda for psychology. *South African Journal of Psychology* 38(2):413-428.

47. Heyer, A. and Ogunbanjo, G.A. (2006). Adherence to HIV antiretroviral therapy Part I: A review of factors that influence adherence. *South African Family Practice*, 48(8):5-9.

48. Roura, M., Busza, J., Wringe, A., Mbata, D., Urassa, M. and Zaba, B. (2009). Barriers to sustaining antiretroviral treatment in Kisesa, Tanzania: A follow-up study to understand attrition from the antiretroviral program. *AIDS Patient Care and STDs* 23(3):203-210.

49. Hardon, A.P., Akurut, D, Comoro, C., and Ekezie, C. (2007). Hunger, waiting time and transport costs: Time to confront challenges to ART adherence in Africa. *AIDS care* 19(5). 658-665.

50. Mukherjee, J.S., Ivers, L., Leandre, F., Farmer, P. and Benforouz, H. (2006). Antiretroviral therapy in resource-poor settings. Decreasing barriers to access and promoting adherence. *Journal of Acquired Immune Deficiency Syndrome* 1 (43): S123-126.

51. Lupu, A.L. (2014). The Role of Acting Participants, Definitions, and the Determining Factors of Adherence to Treatment from Two Perspectives. *Postmodern Openings*, 7(1). 75-88.

52. Simoni, J., Amico, K.R., Pearson, C. and Malow, R. (2009). Overview of adherence to antiretroviral therapies, in *HIV/AIDS: Global frontiers in prevention/intervention,* edited by C. Pope, R.T. White and R. Malow. New York: Routledge.

53. Hogerzeil, H.V.M. (2013). Promotion of access to essential medicines for non-communicable diseases: practical implications of the UN political declaration. *The Lancet,* 381(9867). 680-689.

54. SASSA. South African Social Security Agency. National Department of Social Development. (2011). *you and your grant.* Available at http://www.sassa.gov.za. (Accessed on 10 September 2014).

55. De Paoll, M.M., Mills, E.A. and Gronningsaeter, A.B. (2012). The ARV roll out and the disability grant: A South African dilemma? *Journal of the International AIDS Society* 15(1):6.

56. WHO, World Health Organization, (2006). Interim WHO clinical staging of HIV/AIDS and case definitions for surveillance.

57. AIDSinfomap. 2012. *Community-based support, aids retention, adherence and Treatment response.* Available at http://www.aidsmap.com. (Accessed on 10 July 2014).

58. Yoder, P., Mkhize, S.P. and Nzimande, S. (2009). *Patient Experiences in Antiretroviral Therapy Programmes in KwaZulu-Natal, South Africa.* Durban, South Africa: Health Systems Trust and Calverton, Maryland, USA: Macro International Inc.

59. AIDSinfonet. (2012). *How much adherence is enough?* Available at http://www.aidsinfonet.org/fact_sheets/view/405#FATIGUE (Accessed on 10 July 2014).

60. Ware, N.C. (2009). Explaining adherence success in Sub-Saharan Africa: An ethnographic study: PLoS Med 6(1).